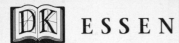
ESSEN

the human genome

JEREMY CHERFAS

SERIES EDITOR JOHN GRIBBIN

London, New York, Munich,
Melbourne, and Delhi

DORLING KINDERSLEY, LONDON

senior editor Peter Frances
senior art editor Vanessa Hamilton
DTP designer Rajen Shah
picture researcher Sarah Duncan
illustrator Richard Tibbitts

category publisher Jonathan Metcalf
managing art editor Phil Ormerod

First American Edition, 2002
02 03 04 05 10 9 8 7 6 5 4 3 2 1
Published in the United States by
DK Publishing, Inc.
95 Madison Ave
New York, NY 10016

A Cataloging-in-Publication record for this title is
available from the Library of Congress
ISBN 0-7894-8414-5

Color reproduction by Mullis Morgan, UK
Printed in Italy by Graphicom

See our complete product line at
www.dk.com

contents

before the genome

On June 26, 2000, the two teams racing to sequence the human genome, to read the recipe for humankind, jointly announced at the White House that they had a first draft. Their declaration underlined that our understanding of how living things reproduce and evolve has come a long way in a short time. Until Charles Darwin wrote *On the Origin of Species,* in 1859, we had no idea about natural selection. In 1953, James Watson and Francis Crick made public the double-helix structure of DNA.

> **" We wish to suggest a structure for the salt of deoxyribose nucleic acid (DNA). This structure has novel features which are of considerable biological interest. "**
>
> James Watson and Francis Crick, 1953

By 1977, the full genetic message of an organism had been read, albeit only a virus. It consisted of just 5,386 letters of genetic code. The first draft of the human genome was published in February 2001. It is about 3 billion letters long.

This book is about the Human Genome Project and what our genome reveals about us. The story begins with the discoveries that made this remarkable scientific project possible.

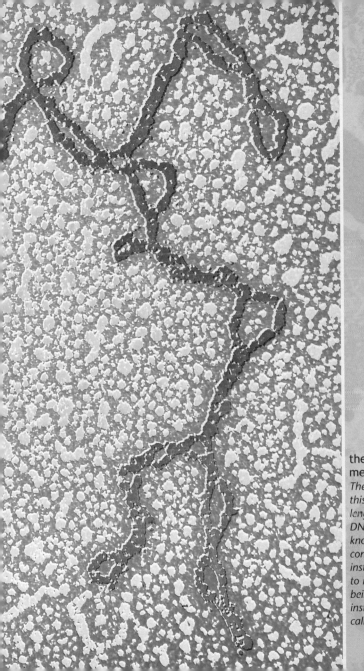

the genetic messenger

The red strand in this picture is a length of human DNA. We now know that our DNA contains the instructions needed to make a human being, the full set of instructions being called the genome.

inheritance and genetics

Science has put a huge effort into explaining something people seem to understand at a very basic level: each child is like its parents. Not identical, but similar enough to let us talk of family likenesses. That, in a nutshell, is heredity. All living things have offspring that are like themselves. That is what enables us to give things a name. What biologists call a species consists of all individuals that are like each other, but distinguishable from a different species. Genetics tells us how the information to build a body passes from one generation to the next. It also answers questions about the origin of species, about evolution.

For most of Western history those questions were simply not asked. Religious orthodoxy said that each species was made by God and did not change. A related doctrine, called transformism, allowed species to change but gave each a separate creation. Finally, there came the theory of evolution, which says that all species alive today trace their ancestors to a single origin of life. Evolution is undeniably true. The puzzle has been to understand how it happens.

There are, really, two questions to answer. Why are offspring like their

order by design
Most religions offer an account of how the universe was created. William Blake's The Ancient of Days *(painted in 1794) encapsulates the Christian belief that the order and regularity observed in the natural world, including living things, are the result of divine creation.*

parents? This is the puzzle of heredity. It enables us to recognize the difference between a dog and an oak. Then there is the question of how one type of organism changes into another. That is the puzzle of evolution. Heredity and evolution are linked because the fact

     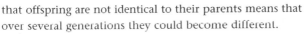

that offspring are not identical to their parents means that over several generations they could become different.

The two linked questions have been answered in two ways, which can be caricatured as the logical and the experimental, genetics and biochemistry. While recognizing that the two strands of inquiry were not only intertwined but also taking place at the same time, it makes the story easier to follow if we separate them, and start with genetics.

ideas about inheritance

Modern science did not invent genetics, although it did give the subject that name, in 1909. In biblical times, people wondered what determined the characteristics of a breed. Jacob, Esau's brother, peeled thin branches so that they were striped along their length, and then put them in the trough where Laban's animals drank. The stock gave birth to young with speckles, streaks, and spots, which Laban had agreed would be Jacob's to keep. So animals can change from generation to generation.

Somewhat different is the idea of inheritance of acquired characteristics,

variation between species
Different species of oak produce acorns that are similar and yet show noticeable differences. Evolution helps to explain this type of relationship between species.

tall story
Writing in 1809, Jean Baptiste de Lamarck put forward the idea that the efforts made by an animal, such as a giraffe, to adapt to its surroundings produce bodily changes that are passed on to its offspring.

often called Lamarckian inheritance after the Frenchman Jean Baptiste de Lamarck. He thought that the things an animal did during its life would be reflected in its offspring. So an animal that stretched to reach leaves at the top of a tree would have young with slightly longer necks. And if they still had to stretch, they in turn would have longer-necked offspring too, until we end up with a modern giraffe. The original ancestor of the giraffe had been separately created, by God, but had also been transformed by the efforts of each generation.

lamarckian inheritance
Lamarck's ideas about inheritance can easily be shown to be false. If inheritance worked in the way he suggested, a dog whose tail had been docked by its owner would eventually give birth to puppies with docked tails.

natural selection

Charles Darwin's genius was to come up with a system – natural selection – that allowed a species to change and give rise to new species. He used three ideas to derive it.

First, all living things are able to have many more offspring than they actually do. Darwin chose to illustrate this with the slowest breeding animal he knew, the elephant, assuming that it starts to breed at 30, stops when it is 90, and has six young over the 60 years. "I have taken some pains to estimate its probable minimum rate of natural increase," Darwin wrote. "After a period of from 740 to 750 years there would be nearly 19 million elephants alive, descended from the first pair."

Second, elephants face environmental pressures, such as a limited supply of resources. The trees and grass they eat cannot multiply as quickly as the elephants. So in every generation some elephants will starve to death.

Finally, in each generation offspring differ slightly from one another. Because they are competing for resources, some inevitably do better than others. They will increase in number, while those that do not do as well decrease.

That is all there is to natural selection. It is the inevitable result of competition between organisms, each slightly different, and heritably so. Darwin's explanation

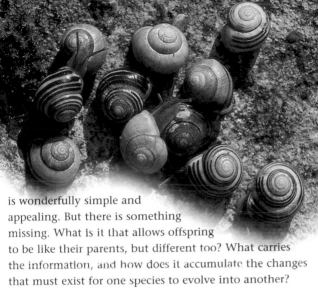

is wonderfully simple and appealing. But there is something missing. What is it that allows offspring to be like their parents, but different too? What carries the information, and how does it accumulate the changes that must exist for one species to evolve into another?

darwin's problem

Darwin himself was not satisfied with the process he had proposed for evolution. It was not enough, he realized, to rely on logic and the evidence of his eyes. He needed an actual mechanism, something that would transmit the information about a species and yet be open to change. Darwin's solution was something he called the "provisional hypothesis of pangenesis." In contrast to the simplicity of evolution by natural selection, this is a complicated set of ideas. Scholars have expended much effort trying to understand Darwin's "provisional hypothesis" but the fact is that it was a dead end. However, in one of the strangest "might-have-beens" of science, the answer was, quite literally, on his desk. Gregor Mendel, normally described as "an obscure Austrian monk," had sent Darwin a copy of his paper, published in 1866. But although this later provided Darwinism with its much-needed mechanism, Darwin himself failed to read it. He was not alone in this. Most of the eminent biologists of the day had access to the paper, and all failed to understand its importance.

mendel's discovery

Gregor Mendel was a gifted scientist who set out to discover how new species arose. Many botanists of the time thought that a hybrid between two existing species might give rise to a new species. Mendel attacked the problem using the garden pea. He studied varieties of pea that differed in single, easily recognized traits. Some were tall, others short; some had white flowers, others purple. In all, Mendel worked with seven pairs of traits.

The pea was an inspired choice because it normally fertilizes itself. Thus Mendel could easily make hybrids between two varieties while knowing that this did not generally happen in nature. Aside from his choice of subject, Mendel differed from his colleagues in focusing on one trait at a time, rather than the overall look of the hybrid, and in the way he meticulously counted the results of his experiments.

The textbook example is of round versus wrinkled peas (see panel, right). Mendel found that one of the two types, in this case wrinkled peas, skipped a generation and then appeared as a quarter of the offspring in the second generation. He found the same pattern in each of the seven traits he studied. Others had observed similar results. Darwin himself crossed common asymmetrical snapdragons with rare, round-flowered types. He found a ratio in the second generation that was slightly lower than

gregor mendel
Gregor Mendel was a monk and, like many monks at that time, a scientist. Born Johann Mendel, he changed his name when he entered the monastery in Brno, north of Vienna, in what is now the Czech Republic.

how traits are inherited

In one of his experiments, Gregor Mendel looked at the way pea shape is inherited. Peas are either round or wrinkled. Mendel said there were two factors for distinct traits like these. We can call them R, for round, and r, for wrinkled. Each plant has two copies of the factors, which we now call alleles. When Mendel crossed two pea plants and allowed the offspring to fertilize themselves (see below), he found that all the peas in the first generation (f1) were round, but that wrinkled peas appeared in the next generation (f2). When Mendel counted the seeds in this generation, he found that the ratio of round to wrinkled peas was 2.96:1

parent plants
Mendel crossed two plants that always bred true: a variety that always produced round seeds (RR) and another that always produced wrinkled seeds (rr).

round pea

wrinkled pea

RR X rr

first generation (f1)
In the first generation, all the plants produced round seeds. The seeds of this generation were then sown and allowed to pollinate themselves.

f1

Rr   Rr   Rr

Rr

all plants in F2 are a cross between Rr and Rr

second generation (f2)
In the next generation, some peas were again wrinkled. It appeared that the wrinkled factor had been dormant for one generation, only to reemerge in the next.

f2

plants with rr have wrinkled seeds

RR   Rr   rR   rr

round   round   round   wrinkled

Mendel's. The breakthrough made by Mendel was to realize that the apparently constant ratio of 3:1 he had noted merited an explanation, and he came up with one. In fact, the modern view of his theory owes a lot to wishful thinking and hindsight, but this is not the place to ask what he really said.

What people think Mendel said is that one of the factors is dominant over the other. This factor is referred to as the dominant allele (R). In the case of round and wrinkled peas, if both the alleles are R, the peas are round. If both are r, the peas are wrinkled. But if there is one R and one r, the peas are round, because R is dominant over r.

Mendel's results encouraged him to formulate three laws of inheritance (see panel, below). Like many laws, however, Mendel's laws have exceptions. There are those who argue that this proves that more than good luck guided Mendel's experiments as he reported them. Perhaps he stopped counting when the results matched his hypothesis or left out results that did not suit his ideas. Frankly, who cares? The exceptions, which apparently did not trouble Mendel, later proved extremely useful as the field he invented matured long after his death.

mendel's laws of inheritance

The Law of Uniformity
When plants that differ in a particular trait are crossed with each other, the offspring are uniform and resemble one parent.

The Law of Segregation of Alleles
The alleles in the parents separate and recombine in the offspring. This explains why the expression of a particular trait in an organism did not resemble a blend between its parents, but resembled either one parent or the other.

The Law of Independent Assortment of Alleles
The alleles of different characteristics pass to the offspring independently, so inheritance of, say, green peas says nothing about whether the peas will be wrinkled or smooth.

mendel rediscovered

Mendel's ratios, and the laws he formulated to account for inheritance, were just what Darwin needed to explain how genetic information is transmitted. Here were factors that were not blended or diluted by breeding but that passed unchanged from one generation to the next. Perhaps natural selection acts on traits less obvious than those Mendel studied, but at least his work pointed the way. For reasons no one can explain, everyone simply ignored him. They did not even bother to prove him wrong.

Vindication came only when Mendel had been dead for more than 15 years. Three men simultaneously and independently did similar experiments and in so doing read, and finally understood, Mendel's original work. They were the Dutch botanist Hugo de Vries, an Austrian, Erich Tschermak von Seysenegg, and a German, Carl Correns.

Make what you will of the fact that these three not only discovered Mendel and replicated many of his results independently, but also published in the same year, 1900. Biologists can only be thankful that they did, although it fell to an English botanist, William Bateson, to beat the drum for Mendel and the science Bateson called genetics.

A professor at Cambridge, Bateson was also an influential member of the Scientific Committee of the Royal Horticultural Society. He discovered the rediscovery of Mendel days before he was due to give a talk to the society. Realizing its importance, he switched the topic of his talk and presented Mendel's paper instead.

change by mutation
Hugo de Vries studied evolution by making crosses between garden varieties of evening primrose plants. Among the offspring he grew, a few were noticeably different from their parents. De Vries coined the term mutation to describe these sudden changes and, in the course of his experiments, devised the same three laws as Mendel.

what are genes?

organism

tissue

cell

nucleus

chromosome

from organism to chromosome
Many living things are organized on several levels. The basic structural component (and the smallest unit in which life can exist independently) is the cell. Similar cells are often organized into tissues. Within the cells of all organisms except bacteria, there is a central nucleus surrounded by fluid called cytoplasm. Among other things, the nucleus houses the chromosomes.

Mendel and his followers looked at the external features of the plants they studied, and used their observations to construct logical explanations of what might be happening inside them. It was clear that discrete particles of hereditary material existed and behaved according to Mendel's laws. In 1909, a Danish botanist, Wilhelm Johannsen, coined the term gene to refer to these particles. But what physical form did genes take?

Some important clues came from another field of biology. Using greatly improved microscopes, made possible by more accurate engineering and manufacturing, biologists had begun to look at living cells. The chemical industry provided dyes that stained some parts of the cell more brightly than others, and some of these dyes were especially useful because they did not kill the cell. Wilhelm von Waldeyer, a German anatomist, noticed in 1888 that the central part of the cell, the nucleus, sometimes contained threadlike bodies that absorbed dyes especially well. He called them chromosomes. Other scientists studied the behavior of these threads as cells divide in two, a process called mitosis (see p.16). They saw that the chromosomes divided first, and that each new cell (known as a daughter cell) inherited a complete set of chromosomes.

During the 1880s, Edouard van Beneden, a Belgian biologist, and Theodor Boveri, a German zoologist, turned their attention to sex cells, or gametes. In all animals, including humans, the gametes are egg and sperm cells, which fuse together during fertilization. Boveri and van Beneden discovered a different pattern to the one seen during mitosis. In this process, known as meiosis (see p.16), cells divide in two stages, eventually producing daughter cells, or gametes, that each contain half the number of chromosomes found in the original cells.

Carl Correns saw the connection between the splitting of the chromosomes as gametes were formed and the segregation of Mendel's second law. Chromosomes behaved, at least at one level, as if they could pass on genetic information as cells divided. But chromosomes turned out to be a complex mixture of different kinds of chemicals, any one of which could have been the hereditary molecule we now know as DNA. The next strand in the story belongs to the biochemists.

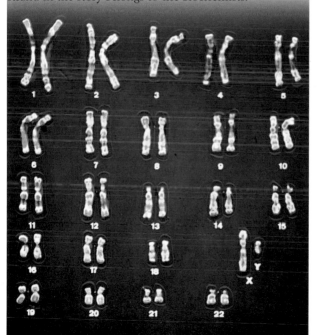

chromosomes
Within the nucleus of a body cell, chromosomes are arranged in pairs. One in each pair comes from the mother, the other from the father. Most human cells contain 46 chromosomes in 23 pairs (shown here). The exceptions are egg and sperm cells (which have 23 single chromosomes) and red blood cells (which have none at all).

mitosis and meiosis

Cells divide and multiply during growth and to replace old cells. As they divide, all the genetic material in the cell is copied in a process called mitosis. Egg and sperm cells arise by a different process, called meiosis, in which each new cell has only half the genetic material of the original.

mitosis

In mitosis, a cell divides to produce two new cells that are identical to the original. For simplicity, only four chromosomes are shown here.

Before division, the chromosomes are duplicated to form x-shaped double chromosomes.

The membrane around the nucleus breaks down, and the chromosomes line up on threads across the cell.

Duplicated chromosomes are separated and pulled to opposite sides of the cell.

Each set of chromosomes is enclosed by a new membrane, and the cell begins to divide in two.

Two new cells are formed, each with a full set of identical chromosomes.

meiosis

The halving of genetic material that occurs during meiosis ensures that a full set of chromosomes will result when sperm and egg cells fuse. As a result of crossing over (stage 2), the genetic material in each new cell is unique.

The chromosomes are copied, forming a set of x-shaped double chromosomes.

The nucleic membrane disappears. Matching chromosomes touch in random places, and genetic material crosses over.

Threads form in the cell and pull the pairs of chromosomes apart. The cells then divide.

The paternal chromosomes (now with some maternal genes) go to one cell; and vice versa.

More threads attach to the chromosomes. The duplicated chromosomes separate into single chromosomes.

The two cells divide to produce four cells from the original cell, each with half the original genetic material.

the molecules of life

It was a young Swiss doctor, Johann Friedrich Miescher, who discovered DNA. Miescher was interested in the chemistry of the cell nucleus. His best source of nuclei was white blood cells, which have large nuclei and which he obtained from pus soaked dressings taken from a local hospital. In 1868, his analysis uncovered a new compound that was acidic and rich in phosphorus, made up of very large molecules. Miescher called it nuclein. A student later suggested the name nucleic acid. Within a few years, its essential chemistry had been worked out.

nucleic acids

The basic unit of nucleic acid was a sugar. Hundreds of these sugars were linked end to end by a phosphate group (a common molecule made of an atom of phosphorus joined to four atoms of oxygen). The third component was a type of compound called a base, attached to the sugar. There seemed to be five different bases, called guanine, adenine, cytosine, thymine, and uracil (usually known by the initials G, A, C, T, and U). A molecule of nucleic acid is made up of subunits, called nucleotides, each consisting of a sugar, base, and phosphate. But while the sugar and phosphate groups alternated predictably, there seemed to be no pattern to the bases.

By the 1920s, it had been discovered that there were two different nucleic acids. One was RNA, found mostly in the cytoplasm (the material that surrounds the cell nucleus). Its sugar was ribose, and it contained the bases

what things are made of
In the early 1800s, chemists realized that living and nonliving things contain the same kinds of atoms and follow the same rules of chemistry. Until then, many thought that living things contained an extra "vital force."

C, G, A, and U; no T. The other, found inside the nucleus, was DNA. Its sugar was deoxyribose, and it contained C, G, A, and T; no U. But, because nucleic acids were so simple, it did not seem possible that they could be the hereditary molecules. There was good evidence that chemicals called proteins play a fundamental role in cells, regulating their activities, passing messages, and acting as structural components. Proteins consist of 20 different subunits, called amino acids. How could a dull nucleic acid with only four kinds of unit carry the information needed to build a protein?

nucleic acids = DNA in nucleus and RNA in cytoplasm

RNA = ribose sugar, phosphate groups, and bases C, G, A, and U

DNA = deoxyribose sugar, phosphate groups, and bases C, G, A, and T

the transforming principle

Oswald Avery, a Canadian bacteriologist, came closest to proving that DNA is the hereditary molecule. His group worked with a bacterium called *Streptococcus pneumoniae*, which causes the fatal disease pneumococcus. Actually, not all the bacteria are virulent. Some strains are harmless and cause no disease. The difference is visible to the naked eye. Growing on an agar plate, the virulent strain makes a smooth, glistening colony, while the harmless mutant makes crinkly, rough-looking colonies. Avery discovered that if he poured killed virulent strain over rough, harmless colonies, some were transformed into smooth colonies. One of the chemicals in the dead virulent bacteria carried

oswald avery
Oswald Avery worked on pneumococcus bacteria at the Rockefeller Institute in New York. By 1944, he and his colleagues, Maclyn McCarty and Colin Macleod, had not only extracted the "transforming principle" that converted rough-coated bacteria into the virulent smooth strain, but had also shown that this was DNA.

the hereditary message that could make harmless bacteria lethal. But which one? Over the next 20 years, Avery and his team devoted themselves to identifying the transforming principle, as they called it. All their tests suggested that this substance was DNA.

back to genetics

Europe led the way in the new science of genetics until the start of the 20th century, when the focus shifted decisively to the United States and the laboratory of one man, Thomas Hunt Morgan.

Morgan's discoveries would bring together genetics, evolution, and development, the process by which a single cell gives rise to an entire animal. He and his students established three things about genes, which they could map to specific locations along the chromosomes. First, genes were the units of Mendelian inheritance, the factors that determined traits. Second, different forms of genes provided the variability that evolution by natural selection needed. And, finally, some genes were the switches that controlled development.

Morgan arrived at Columbia University in New York in 1904 determined to understand development. He found that the fruit fly made an ideal subject for his experiments. Morgan and his group patiently bred flies until a mutation, a sudden change, appeared. The mutant was a male with white eyes, unlike wild flies which have red eyes.

When the white-eyed male was bred with a red-eyed female, all their offspring, the F1 (first) generation, had red eyes. So the allele for red eyes is dominant over the allele for white eyes. Morgan crossed brother and sister from the F1 generation to breed the F2 (second), and obtained the expected, Mendelian, result: three red-eyed flies to every white-eyed fly. But there was something unexpected, too.

Only males had white eyes. All the females, and some of the males, had red eyes. The fruit flies had broken Mendel's third law; the gene for white eyes was linked to the gene for sex. Looking at the chromosomes, Morgan could see that three pairs looked the same in males and females. But

flying start
On finding that mice and rats bred too slowly for his purposes, Morgan turned to the fruit fly, Drosophila melanogaster. It is small – a thousand or more flies can fit into a small bottle. It breeds quickly, producing 30 generations a year. It is easy to observe, as its eggs develop outside the female's body. And it is simple, having only four pairs of chromosomes.

sex-linked inheritance

Morgan had worked out that a male gets the X chromosome from its mother and the Y from its father. A female gets one X chromosome from each parent. He hypothesized that the gene for eye color is on the X chromosome, which also carries the gene for sex. There is no eye-color gene on the Y chromosome, so in male offspring whichever gene is on the X chromosome will be expressed.

X and Y: chromosomes
W and w: genes

female WW

If the mother has two red-eye genes (shown here as W), all her sons will have red eyes, even if their father has white eyes.

♀ X X
  W W

♂ X Y
  w

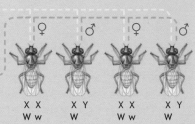

♀   ♂   ♀   ♂

X X   X Y   X X   X Y
W w   W   W w   W

female Ww

If the mother has one red- and one white-eye (w) gene, half her sons will have white eyes and half red, no matter what the color of their father's eyes.

♀ X X
  W w

♂ X Y
  W

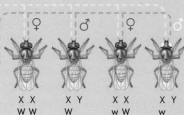

♀   ♂   ♀   ♂

X X   X Y   X X   X Y
W W   W   w W   w

female ww

If the mother is white-eyed and the father is red-eyed, all sons will have white eyes and daughters will have red, as the father's X chromosome has the dominant red allele.

♀ X X
  w w

♂ X Y
  W

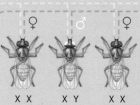
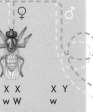

♀   ♂   ♀   ♂

X X   X Y   X X   X Y
w W   w   w W   w

♀ red-eyed female   ♀ white-eyed female   ♂ red-eyed male   ♂ white-eyed male

the fourth pair differed. Females had a pair of matched chromosomes, called the X chromosomes. Males had a single X chromosome that paired with a much smaller Y chromosome, which is never found in females. Morgan proposed that the gene for eye color is linked to the gene for sex (see panel, left). He concluded that genes reside on chromosomes, and that each gene resides at a particular place, called its locus, on a particular chromosome.

recombination

Next came a huge conceptual leap. The chromosome is a long line of genes. Genes physically close to one another tended to stay together in subsequent generations. The traits were linked, just like eye color and sex. Very occasionally, however, linked traits would indeed separate. From this, Morgan developed the idea of recombination (or crossing over). During meiosis, the paired chromosomes come together before separating into the daughter cells. Morgan thought that parts of strands might be swapped, so that a part of the mother's chromosome and a part of the father's chromosome came together to create an entirely new combination of genes in the offspring.

Here was the physical, biochemical basis of that crucial observation, that offspring are somewhat like both of their parents. Of course they are; they have some of their father's genes and some of their mother's, and each combination (apart from that in identical twins) is unique. Morgan then went further. If the genes were strung out along the chromosome like beads, then the chances of two beads being separated during crossing over depended on how far apart they were. And if they were on different chromosomes, different necklaces, they would always be inherited independently, just as Mendel's law said they ought to be.

a genetic necklace
Genes are arranged in order, like beads on a necklace. The chances of two genes on the same chromosome (such as the selected fruit-fly genes shown here) being inherited together depend on the distance between them. Genes that are close together will usually be inherited together; those that are far apart will almost inevitably be separated.

hairy body

scarlet eyes

curled wings

spineless bristles

striped body

delta veins

ebony body

rough eye surface

chromosome

band produced
by staining

locating genes
When stained, the chromosomes of the fruit fly can be divided into the light and dark bands seen here. Calvin Bridges was able to assign genes to bands like these, creating one of the first physical maps of genetic information.

stamp of approval
This Swedish postage stamp commemorates the Nobel prize awarded to Thomas Hunt Morgan in 1933. Morgan was the first native-born American to win the prize.

Morgan's offspring

A student of Morgan, Alfred Henry Sturtevant, in one feverish night in 1911, "to the neglect of my undergraduate homework," as he later recalled, took some data on inheritance patterns (or linkage) in *Drosophila* and created the first map of the five known genes on its X chromosome. His map showed the order of the genes and how likely they are to stay together during recombination. Around the same time, another of Morgan's students, Calvin Bridges, exploited an unforeseen advantage of the fruit fly to create a physical map that corresponded to Sturtevant's linkage map. The cells of the fly's salivary glands contain giant chromosomes that, when stained, show distinctive patterns of light and dark bands. Close study of the bands revealed physical changes that mirrored the genetic changes. When Morgan won the Nobel prize in 1933, he generously shared the prize money with Sturtevant and Bridges.

Hermann J. Muller discovered that X-rays dramatically increase the rate of mutation. That made it possible to study genetics, inheritance, and development much more rapidly. Muller won the Nobel prize in 1946. George Beadle, who trained with Morgan, worked with Edward Tatum and used X-rays to create mutations in the mold *Neurospora*. By tracing the effects of mutations, they discovered that genes carry the code to make proteins, many of which are enzymes – chemicals that enable and control reactions in cells. Beadle and Tatum summarized their work as "one gene, one enzyme." They shared a Nobel prize in 1958 with another Columbia student, Joshua Lederberg.

the double helix

By the beginning of the 1950s biologists knew that DNA carried the hereditary message. But how?

It had to be something to do with the way the DNA molecule itself was organized. In a flurry of activity in the spring of 1953, the structure of DNA was worked out by the Englishman Francis Crick and the American James Watson. Their one crucial breakthrough was to realize that the four kinds of bases (C, G, A, and T) fit together like jigsaw puzzle pieces (see panel on p.24). The beauty of the structure of the molecule is that it allows DNA to transmit information from generation to generation with great simplicity, because each of the two helixes contains all the information needed to build its opposite half.

code breakers

The structure of DNA answered the question of how DNA could pass information from generation to generation, but not about how that information was actually carried on the molecule. The order of amino acids in a protein determines its structure and properties, and somehow the DNA stores that order. A code must translate the order of bases along the DNA into the order of amino acids along a protein; labs around the world set out to crack the genetic code.

Logically, the code had to use at least three bases to represent each amino acid. The code for an amino acid can be thought of as a word made up of letters corresponding

mapping a molecule

An English physicist, Rosalind Franklin, uncovered some vital clues about the structure of the DNA molecule. She used a technique called X-ray crystallography, in which she passed X-rays through DNA in a crystal-like form. The shape of the shadow suggested that the molecule was a helix.

model answer

In their pursuit of the structure of DNA, Crick and Watson used models of the components. They knew, from the evidence of Franklin and others, what shape the DNA should be. Once they had worked out how the bases joined together, they were able to build a model of the double helix.

the structure of DNA

The DNA molecule can be visualized as a kind of spiral ladder. The ladder's sides consist of smaller sugar and phosphate molecules; the rungs are formed by base molecules, which occur in pairs. Genetic information is represented by these base pairs – a sequence of base pairs that contains the information needed to make a single protein is a gene. Bases join together in such a way that each half of the helix holds the information needed to build its opposite half (known as its complement), so that when the two strands divide, a perfect copy of the molecule can be made.

tightly coiled strand of DNA

the DNA molecule

A chromosome consists of a long, thin DNA molecule surrounded by protein. In its normal state, DNA is spread out in the cell nucleus, but just before cell division it packs into the compact structure shown here.

a gene is a length of DNA, usually several thousand bases long

"rungs" formed by base pairs

two strands coil around each other to form double helix

how bases join together

The four bases in a DNA molecule pair in a specific way. Adenine and guanine are large and are called purines. Thymine and cytosine are small and are called pyrimidines. A will join to T (but not to C) and G will join to C (but not to T), and when they do the shape of the base-pair is identical in both cases. The only difference is that A and T join at two places while G and C join at three.

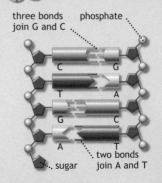

three bonds join G and C

phosphate

two bonds join A and T

sugar

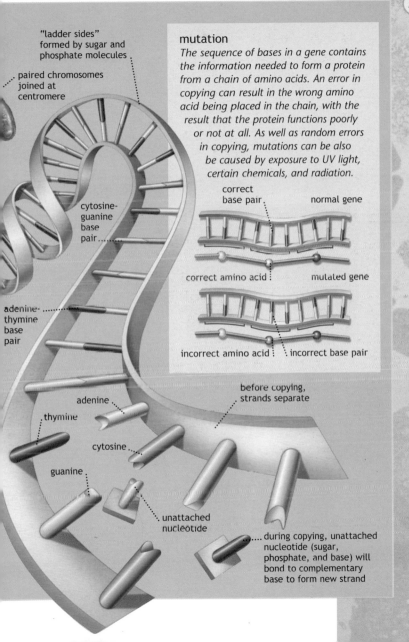

"ladder sides" formed by sugar and phosphate molecules

paired chromosomes joined at centromere

cytosine-guanine base pair

adenine-thymine base pair

mutation

The sequence of bases in a gene contains the information needed to form a protein from a chain of amino acids. An error in copying can result in the wrong amino acid being placed in the chain, with the result that the protein functions poorly or not at all. As well as random errors in copying, mutations can be also be caused by exposure to UV light, certain chemicals, and radiation.

correct base pair

normal gene

correct amino acid

mutated gene

incorrect amino acid

incorrect base pair

adenine

thymine

cytosine

guanine

unattached nucleotide

before copying, strands separate

during copying, unattached nucleotide (sugar, phosphate, and base) will bond to complementary base to form new strand

from DNA to protein

Proteins play various roles in the body: some form structures such as skin or hair; others, such as enzymes and hormones, control cell activity. The translation from gene to protein starts in the cell nucleus, where a copy of the gene is made. The copy, called a messenger RNA (mRNA) strand, then moves to the cytoplasm, where it is converted to a protein by an assembly unit called a ribosome and special adaptor molecules of transfer RNA (tRNA), which bring different amino acids into place.

DNA strands separate

mRNA strand made from bases in nucleus

unattached nucleotide

cell nucleus

transcription

The helix unwinds and the DNA strands separate. Using one half as a template, a molecule of mRNA is assembled from unattached bases in the nucleus. The mRNA then breaks away from the DNA template and moves into the cytoplasm.

mRNA strand

ribosome moves along mRNA strand, reading codons

tRNA molecule delivers amino acid

amino acid

codon on tRNA complements codon on mRNA

assembling a protein

A ribosome moves along the mRNA strand. As it reads each codon, the corresponding amino acid is delivered by a molecule of tRNA and added to a growing chain. The tRNA molecule disengages and the ribosome advances to the next codon.

amino acid added to chain

tRNA disengages

enzyme

amino acid chain

assembly complete

On reaching a stop codon, the ribosome falls away and the amino-acid chain is released. The chain then folds up to form the completed protein.

chain twists and folds into finished protein

to the bases. With four letters, it is possible to produce 16 different two-letter words. If the words are three letters long, there are 64 possible words, which is enough. Four-letter words would be overkill. Scientists set about working out which three-letter words (also called codons or triplets) code for which amino acids. Within a matter of years, all 64 codons had been translated: 60 code for amino acids, while three are stop signals and one is a start signal.

The code proved to be virtually identical in all living things examined – powerful evidence that all life on Earth had descended from a single common ancestor.

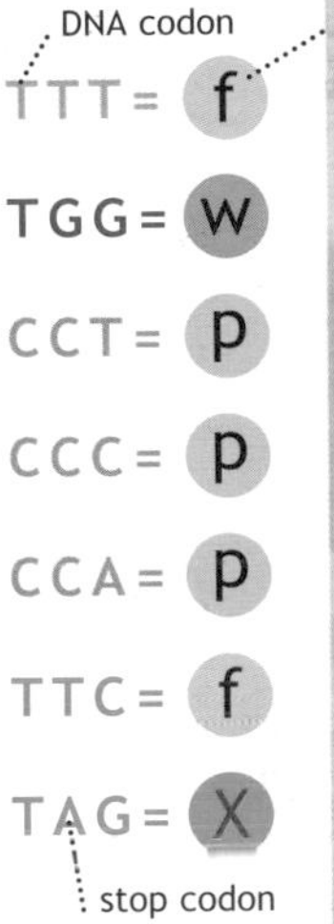

DNA codon · · · · · · · · · · amino acid

TTT = **f**

TGG = **w**

CCT = **p**

CCC = **p**

CCA = **p**

TTC = **f**

TAG = **X**
stop codon

coding for amino acids
The 64 different three-letter DNA codons translate into 20 amino acids. A sample group of amino acids (represented by standard single-letter abbreviations) and their corresponding codons are shown here. Most amino acids are coded for by more than one codon, while growing protein chains are terminated by one of three stop codons.

on the eve of a revolution

By the end of the 1960s, molecular biology seemed to have been pretty much figured out. Genes code for proteins and are stretched out in linear fashion along the chromosomes. The hereditary message is carried by nucleic acids – DNA in most forms of life, RNA in a few viruses. It is carried as a triplet code, in which a word of three bases specifies a particular amino acid. The base sequence is used to manufacture proteins within cells by a process called translation (see panel, left). Exactly how all this works is a matter of detail, and around the world molecular biologists were engaged in the useful if dull work of finding out more and more about less and less.

❝ The two chains of the DNA, which fit together as a hand fits a glove, are separated in some way and the hand then acts as a mould for the formation of a new glove while the glove acts as a mould for a new hand. Thus we finish up with two gloved hands where we had only one before. ❞

Francis Crick, 1957

reading the genes

From the moment molecular biologists
deciphered the first triplet of the genetic
code, they were enthralled by the idea of reading
the entire genetic message of an organism, its
genome. It would, in a very real sense, be the book
of life, and the thought of being able to look in it
for clues to particular mysteries was appealing. But
early efforts to read the genome were tedious and
boring. In fact, when the idea of deciphering the
human genome was first discussed, Sidney
Brenner, who had been instrumental in making
sense of the role of messenger RNA, said, not
entirely in jest, that the work was so spirit-
crushingly dreary that it ought to be the work of
convicts. All that changed, more rapidly than
proponents of sequencing dared hope. The
changes, however, were not so much in the detail
of how to sequence, but in the machinery
brought to bear on the problem. With a couple of
crucial exceptions, the fundamental concepts of
reading DNA sequences have not altered since the
first genome was sequenced in the late 1970s.

```
GGGAGGCTGCTCCTTTTCCTCCGAAAGTCT
AAAGGGAGCGCATTGAGGCCCAGAATAGG
CCTTTGATGCCAAAACATCTGTCTTTTGTG
CGGAGCCCAAAGAATCCTTTGTCAAAGGG
CCATCCAGAGCAGAGAAGGAGGAAAAGTG
CGGTGAAGACTGAGGGAGGGAGCGACTCTG
CAGTGAAGGATGATCAGGTCTTCCCCATG
ACCCTCCCAAATATGACAAGATCGAGGA
TGGCCATGATGACTCATCTGCATGAGCCT
CTGTGCTGTACAACCTCAAAGAACGTTAT
CAGCCTGGATGATCTACACCTATTCAGGT
TCTTCTGTGTCACTGTCAACCCCTACAAGT
GCTGCCTGTGTATAAGCCCGAGGTGGTGA
AGCCTACCGAGGCAAAAAGCGCCAGGAGG
CCCGCCCCACATCTTCTCCATCTCTGACAA
GCCTATCAGTTCATGCTGACTGACCGAGA
AATCAGTCAATCCTGATCGCTGGAGAATC
GGTGCAGGGAAGACTGTGAACACCAAGCG
GTCATCCAGTACTTTGCAACAATTGCAGT
ACTGGTGAGAAGAAGAAGGAAGAAATTAC
TCTGGCAAAATACAGGGGACTCTGGAAGA
CAAATCATCAGTGCCAACCCCCTACTGGA
GCCTTTGGCAACGCCAAGACCGTGAGGAA
GACAACTCCTCTCGCTTTGGTAAAATTCAT
AGAATCCACTTGGCACTACTGGAAAACT
GCATCTGCTGATATTGAAACATATCTGCT
GAGAAGTCTAGAGTTGTTTTTCCAGCTTAA
GCTGAGAGAAGTTATCATATTTTTTACCA
ATTACATCGAATAAGAAACCAGAACTTAT
GAAATGCTTCTGATTACCACGAACCCATA
GATTACCCATTTGTCAGTCAAGGGGAGAT
AGTGTGGCCAGCATCGATGATCAGGAAGA
CTGATGGCCACAGATAGTGCTATTGATAT
TTGGGCTTTTACTAATGAAGAAAAGGTCTC
ATTTACAAGCTCACGGGGGCTGTGATGCA
TATGGGAACCTAAAATTTAAGCAAAAGCA
CGTGAGGAGCAAGCAGAGCCAGATGGCAC
GAAGTTTGCTGACAAGGCGGCCTACCTCCA
AGTCTGAACTCTGCAGATCTGCTCAAAGC
CTCTGCTACCCCAGGGTCAAGGTCGGCAA
GAGTATGTCACCAAAGGCCAGACTGTAGA
CAGGTGTCCAACGCAGTAGGTGCTCTGGC
AAAGCCGTCTACGAGAAGATGTTCCTGTG
ATGGTTGCCCGCATCAACCAGCAGCTGGA
ACCAAGCAGCCCAGGCAGTACTTCATCGG
```

chromosome 17
This is a very small part of the DNA sequence on human chromosome 17. The sequence of this chromosome was read at laboratories in the United States and Germany. In all, 20 laboratories in six countries were involved in the Human Genome Project.

a toolbox

DNA scissors
The DNA molecule is extremely long. To read its sequence, gene mappers need to break it into more manageable pieces. Some of the most useful tools for cutting DNA are chemicals called restriction enzymes, which react with DNA and break it at specific sequences.

sticky ends
The restriction enzyme EcoRI recognizes the sequence GAATTC on one strand, which will be opposite CTTAAG on the other. The DNA is cut between the G and the A on both strands, leaving a single strand of TTAA, known as a sticky end, projecting at each end of the fragment.

Sequencing a genome involves reading the nucleotide bases as they occur along the DNA molecule. But this is not the whole story. For the sequence to be meaningful, gene mappers need to find out where each gene begins and ends. They also need to know what each gene does.

As with so many technological advances, speedy methods for sequencing depended on the development of better tools. Although sequencing today makes use of a vast toolkit, two tools in particular are indispensable: precision "scissors" to cut DNA, and equipment and techniques for separating fragments produced by cutting.

The scissors are enzymes found in all sorts of bacteria. They are called restriction enzymes because they restrict the ability of viruses to grow in the bacteria, by chopping the virus's DNA into small pieces. Mechanical methods, such as violent stirring or ultrasonic vibrations, will also fragment DNA, but the breaks occur at random. Restriction enzymes are much more specific, so each time they are used they break the DNA at exactly the same sequence.

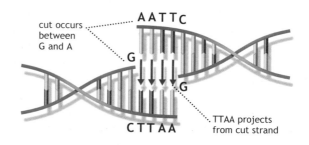

cut occurs between G and A

A A T T C

G

G

C T T A A

TTAA projects from cut strand

separating fragments

DNA fragments can be separated from one another using a technique called gel electrophoresis. A mixture of fragments is loaded at one end of a gel made of long molecules loosely linked together. DNA fragments are electrically charged, so when a current is applied they tend to move through the gel. As the fragments move, the gel molecules hinder their progress. Small fragments slip through easily, and so move faster than larger ones. Over time, the mixture of fragments spreads out through the gel.

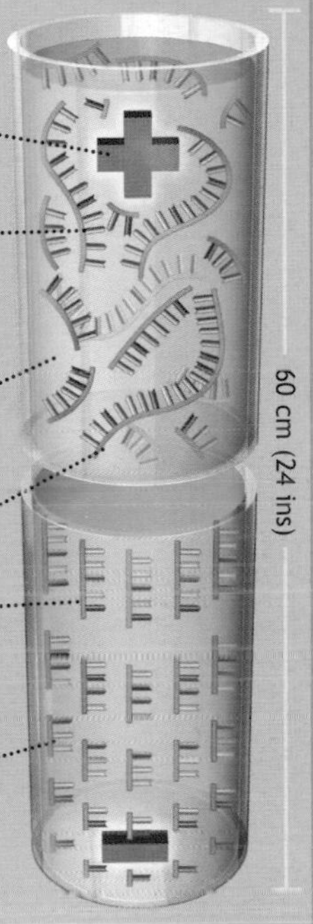

gel electrophoresis
In gel electrophoresis, DNA strands move across a flat plate or through a capillary tube (shown here).

electric current applied to gel

fragments of different length loaded at one end of tube

gel of long, linked molecules

fragments move along gel

long fragments move more slowly than short ones

identical fragments accumulate in separate bands

60 cm (24 ins)

Each restriction enzyme recognizes a particular genetic sequence (called a recognition site), and cuts the DNA at that sequence and nowhere else. The enzymes are named after the bacteria in which they were first found. Thus, the first enzyme isolated from *Escherichia coli* is known as *Eco*RI. When *Eco*RI cuts DNA, it creates what is known as a sticky end (see left). Other restriction enzymes cut straight across the double helix, so the ends are flush, but sticky ends are particularly useful because they allow two sets of DNA, perhaps from different organisms, to be joined. If DNA strands from two organisms are mixed and cut with *Eco*RI, then all the fragments will have the same sticky ends. These sticky complementary strands attract each other, and so a fragment from one organism is likely to end up joined to a fragment from the other organism. This is the basis of the technique by which molecular biologists stitch pieces of DNA together, combining DNA from different sources to make a recombinant DNA molecule.

Different restriction enzymes recognize different sequences, but the crucial point is that a given restriction enzyme will always cut a given piece of DNA into the same set of fragments. That makes it possible to obtain many identical fragments from a sample of DNA.

The second tool crucial for sequencing allows fragments of DNA to be separated from one another and arranged in groups of identical fragments. This technique is known as gel electrophoresis (see panel on previous page).

the first sequences

With restriction enzymes and gel electrophoresis (as well as with other tools) the method used to read a base sequence is not hard to grasp. Start with a purified sample of DNA. Break it with a restriction enzyme. This gives an identical starting point for all fragments. Now make multiple copies of the fragments but, by stopping their growth at random, ensure that these copies are incomplete. This gives a set of fragments that all start at the same point but end a different number of bases away. Ideally, there will be some fragments one base longer than the start, some two bases longer, and so on, up to fragments a few hundred bases long. During copying, the base at the end of each fragment is labeled with fluorescent dye. Once the strands have been

broken strands
To read the base sequence of a DNA fragment, randomly terminated copies of the fragment, each starting with the same letter, are first arranged in order of length. By identifying the last letter of each fragment (reading from the bottom up in the example shown here), the sequence of the original strand can be read.

AATCGGA
AATCGG
AATCG
AATC
AAT · · · bases read from fragment ends match original sequence
AA
A · · · original sequence
AATCGGA

sorted by gel electrophoresis, the sequence of the original strand can be found by reading the labeled base at the end of each fragment with a laser (see left).

There are two methods for sequencing. One was devised by Fred Sanger of Cambridge University, England, and is the basis of modern sequencing. The other, devised by Walter Gilbert and Allan Maxam at Harvard University, is no longer used. Sanger and Gilbert shared half the 1980 Nobel Prize for Chemistry. In fact, the details of the methods do not matter, since only sequencers need to understand them – and, anyway, today the work is done by machines. What does matter is that, whichever method a scientist chose, he or she could sequence genes rapidly and reliably.

The new methods quickly revealed some unexpected surprises, the greatest of which was that genes were not the simple linear stretches of DNA people had imagined. The average gene consisted of several sequences that coded for protein but which were buried in long stretches that seemed not to code for anything. One example was the gene for collagen, a protein that is a key component of ligaments and tendons. The "gene" is about 38,000 bases long. But the coding sequences cover just 5,000 bases spread out over the 38,000 bases and interrupted by more than 50 stretches of "nonsense" sequence.

THEAADEGJI**GENE**WISPX**AB**W**OUT**KQHKS**THIRTY** SGIU**EIGHT**KPAKW**THOU**ASL**SAND**L**BASES**AEESSF **LONG.**FBUT**K**THE**SXHE**COD**XING**WR**SEQUENCES** DHFF**COVER**ASEWD**JUST**KGFL**FIVE**Q**THOUSAND** POIJVVI**BASES**TYD**SPREAD**DFGUE**OUT**JSOEFJWEJF FVLFFF**OVER**YOA**THE**XE**THIRTY**WDJODG**EIGHT**P **THOUSAND**YFIDD**BASES**JHCCS**AND**YAOCKR,P**IN TERRUPTED**SSBYM**MORE**TGEDS**THAN**WAVFKPA SCS**FIFTY**HDQA**NONSENSE**SOAID**SEQUI**JV**ENCES.**

frederick sanger
The English scientist Frederick Sanger is one of only two scientists to have won the Nobel prize twice in the same discipline. He was first awarded the prize in 1953, for his work on the structure of the protein insulin. In 1980, he earned a second prize for his work on gene sequencing, studying a virus called PHI X1/4, which affects the bacteria E. coli.

junk DNA
The meaningful part of the message of DNA – the part that contains the code for proteins – is interspersed with letters that appear to serve no purpose. These apparently meaningless letters are sometimes called junk DNA.

the human genome project

The idea of sequencing the whole human genome first arose in the mind of an American called Robert Sinsheimer. In 1985 he assembled some of the brightest minds in the young field of gene mapping to mull over the idea. They agreed that it was bold and exciting – but just not realistic. At the time, they couldn't even contemplate sequencing a bacterium. The largest organism then sequenced was a virus that has only 150,000 bases. Starting work on the human genome seemed utterly insane.

A few people at that meeting, notably Walter Gilbert, were enthusiastic about the idea. Gilbert won James Watson (who had played such a key part in discovering the structure of DNA) over to the sequencers' side. Slowly, other scientists, some of them powerful administrators of scientific institutions, joined the cause. Within a year, feelings had shifted: it was possible. But a new objection appeared: although it was possible, it should not be done because it would take money from

Epstein-Barr virus genome (150,000 bases)

human genome (3,000,000,000 bases)

whose sequence?

Early on in the Human Genome Project, many people wondered just whose sequence would be read. Every individual's genome differs from everyone else's (by about one letter in 500, we know now, thanks to the sequencing effort). So who would represent the canonical human genome? There was a rumor that it was James Watson himself who was being sequenced. It doesn't really matter, because at every place where individuals differ, the sequence is annotated to identify the differences. And each new difference is added to what is known as the consensus sequence. So the final sequence is really both nobody's and everybody's.

more pressing research problems. And in any case, why bother with the whole sequence, most of which was "junk"? Why not just sequence the genes, the pieces that mattered, as and when they were needed?

the ball starts rolling

Over the next couple of years, the arguments raged back and forth, as did power struggles over funding and control of the project. By 1988, the National Institutes of Health (NIH) in the United States had created an office for genome research, with Watson as its director. Britain's Wellcome Trust, a huge medical charity, came on board, as did other large laboratories around the world. The Human Genome Project was underway, with a carefully thought-out strategy.

First, they would not start with the human genome; nor would they start with sequencing. Instead they decided to build detailed maps of the chromosomes, and to start with the bacterium *Escherichia coli*, the nematode worm *Caenorhabditis elegans*, yeast, and eventually the house mouse *Mus musculus*. Maps are useful in their own right, for people hunting individual genes, and would make it easier for sequencers to position their results on the

amplifying DNA

Scientists working on genome sequences need to make multiple copies of fragments of DNA. This is called DNA amplification. In the late 1980s, a new technique, called the polymerase chain reaction (PCR), made it possible to amplify DNA far more quickly than had been possible before. PCR uses an enzyme called polymerase, which can manufacture complementary DNA in a test tube, given a single strand as a template and a supply of the four nucleotides (A, T, C, and G). It also needs two short DNA sequences, called primers, which complement regions on either side of the template. The reaction is run in a series of cycles, one of which is shown here. With each cycle, the amount of DNA is doubled.

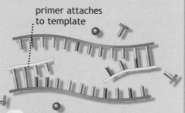

the reaction mixture
The template strand, polymerase, nucleotides, and primers are put in a tube and placed in a machine that regulates its temperature.

strands separate
The mixture is heated to 203°F (95°C), causing the DNA strands to separate. The mixture cools to 99°F (37°C), and the primers attach to their complementary sequences on either side of the template.

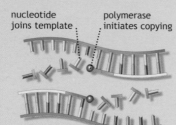

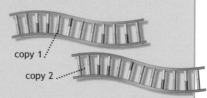

nucleotides attach
When the mixture is heated to 162°F (72°C), the polymerase makes a new strand starting from the primer and working along the template.

two copies
At the end of the cycle, there are two copies of the original template sequence. After another cycle, there will be four copies of the template.

genome once technology had become cheap enough and quick enough for work on the actual sequence to begin.

There are two kinds of map: genetic and physical maps. On a genetic map, markers are placed in the correct order and position relative to each other on the chromosomes. T.H. Morgan's first linkage map was a genetic map. When his colleagues identified genetic mutations with visible patterns on chromosomes, they made the first physical map, on which markers are placed at their actual positions on the chromosomes. The sequence of every base is the ultimate physical map.

While the early maps were being made, there was further progress on developing tools for handling DNA. One of the most useful of these proved to be a method for making artificial copies of DNA fragments (see panel, left).

genetic map — physical map

- genes and markers can be located on both maps
- INSR
- gene for insulin-resistant diabetes (INSR)
- regions of chromosome identified by chemical staining
- gene for hemolytic anemia (GPI)
- DM
- gene for myotonic dystrophy (DM)

21.5
10.8
9.9
10.5
17.9
18

genetic and physical maps
On a genetic map, distances between genes and markers (markers are shown here by flags) are measured in "recombination frequency" (marked down the left of the map). This is the probability that the genes and markers will separate during meiosis. On a physical map, the same features can be assigned to regions identified by staining. Distances on this type of map can be measured in base pairs.

sequencing signposts

A genetic map needs markers, and lots of them; a map with a widely spaced grid is hard to use when you want to find a street, rather than a country. The first markers were visible characteristics, such as the white-eye gene. But any difference between individuals, known as a polymorphism, can be a marker. Imagine a single mutation that changes the recognition site of a restriction enzyme. The enzyme will no longer cut there, so there will be one fewer fragment, which will show up when the fragments are separated on a gel. That kind of marker is called a restriction fragment length polymorphism (or RFLP). Lots of RFLPs and other kinds of polymorphisms have no relationship to specific genes. But they can be detected and their inheritance followed, and that is all that matters. These markers were to play a vital role, because as more were identified and the gaps between them became smaller, they could be used to identify regions of the chromosomes that were small enough to be sequenced (see panel, below).

By 1990, maps relating genetic markers to the physical chromosomes were almost complete for the nematode *Caenorhabditis* and yeast. Sequencing technology had

making the map

A sequencing machine can read only about 500 bases at a time. This means that the whole sequence has to be assembled from hundreds of thousands of small fragments. Putting the fragments in order is the hard part. In practice, mappers start with both the whole chromosome and small fragments. A method known as top-down mapping is used to make a low-resolution chromosome map, which becomes more detailed as more markers are given their correct place on the map. A complementary technique called bottom-up mapping begins with a stretch of DNA with markers that locate it on the top-down map. This is broken into small pieces, which are put in order by examining their own markers. The mappers then decide which pieces to sequence in detail. The final sequence is produced by combining the top-down and bottom-up maps.

top-down map
The chromosome is cut into large pieces. Markers are assigned in several stages to make a progressively more detailed map.

BAC.

bottom-up, stage 1
The chromosome is broken at random into large fragments. Each fragment is cloned using bacteria, to make a bacterial artificial chromosome (BAC).

improved, and trials to see how feasible it was to sequence long stretches of DNA were under way. The various human chromosomes had been assigned to the centers taking part in the project, individual important genes were being found and sequenced, and everything seemed to be going according to plan.

At that point Craig Venter, a scientist who worked at the National Institutes of Health, which is funded by the US government, threw a monkey wrench into the works.

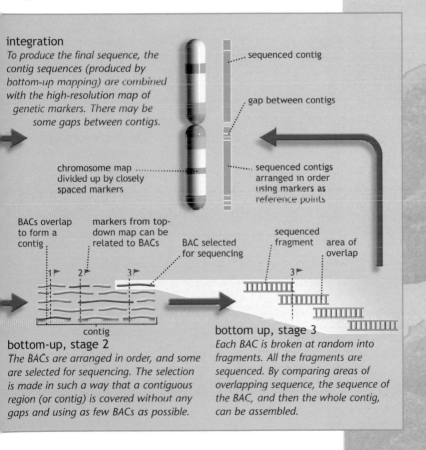

integration
To produce the final sequence, the contig sequences (produced by bottom-up mapping) are combined with the high-resolution map of genetic markers. There may be some gaps between contigs.

sequenced contig

gap between contigs

chromosome map divided up by closely spaced markers

sequenced contigs arranged in order using markers as reference points

BACs overlap to form a contig

markers from top-down map can be related to BACs

BAC selected for sequencing

sequenced fragment

area of overlap

contig

bottom-up, stage 2
The BACs are arranged in order, and some are selected for sequencing. The selection is made in such a way that a contiguous region (or contig) is covered without any gaps and using as few BACs as possible.

bottom up, stage 3
Each BAC is broken at random into fragments. All the fragments are sequenced. By comparing areas of overlapping sequence, the sequence of the BAC, and then the whole contig, can be assembled.

a two-horse race

Craig Venter and his colleague Mark Adams adopted a technique that made use of DNA fragments known as expressed sequence tags, or ESTs, which found genes at record speed, a thousand a month. They didn't know what the genes coded for, but the sequences were likely to be genuine coding stretches of DNA, rather than junk. So why not sequence just the genes and leave the junk till later?

ESTs use the body's own protein-making machinery to find the genes. Instead of breaking the whole genome, junk and all, into parts, Venter's group made copies of all the messenger RNA (mRNA) they could find in various types of cell. From the mRNA, they could make a DNA copy and then sequence that. The copy was of DNA that had been expressed (that is, it had been translated into mRNA), hence "expressed sequence tag." And with the EST they could go fishing in the whole genome to find where, on the chromosomes, the gene that produced the EST was located (see right).

FISHing for genes
Craig Venter and his colleagues used a technique called fluorescence in-situ hybridization (FISH) to find the location of fragments of DNA that match expressed sequence tags. In this method, a tag is labeled with a fluorescent marker and inserted into a cell, where it attaches to the complementary region on the chromosome. That region can then be identified by looking for the glow of the marker.

tag located by fluorescent marker

whole-genome shotgun sequencing

Venter left the National Institutes of Health and set up two private companies. One, The Institute for Genomic Research (TIGR), was to be a nonprofit company that used the latest machinery to sequence genomes. The other, Human

Human Genome Sciences (HGS), would have access to TIGR's data and use it to make a profit. With access to fast sequencing machines and powerful computers, Venter decided to use an alternative approach to sequencing.

This method, developed by Fred Sanger in 1982, is called whole-genome shotgun sequencing, because at the start of the process a chromosome is fragmented at random, as though a cell-sized shotgun has been fired at the nucleus (see panel, below). TIGR set its sights on *Haemophilus influenzae*, from which one of the first restriction enzymes had been isolated. The *H. influenzae* genome is smaller than that of *E. coli*, which after nine years of government funding was still incomplete. In 1994, Venter applied to the NIH for a grant to test his new method by sequencing *H. influenzae* but he began work without waiting for a reply.

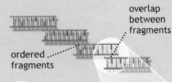

an early success
The bacterium Haemophilus influenzae *causes several infections in humans, including meningitis and pneumonia. Having sequence data for microorganisms has added greatly to biologists' understanding of how they function.*

shotgun sequencing

Shotgun sequencing is a relatively simple method of reading a genome sequence because it does away with the need to locate individual DNA fragments on a map before they are sequenced. The method relies on powerful computers to assemble the finished sequence.

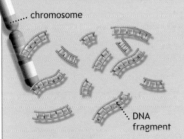

.... chromosome

.... DNA fragment

fragmentation
A chromosome is randomly shredded into thousands of small fragments, which are then sequenced.

overlap between fragments

ordered fragments

assembly
A computer is used to compare the sequences of the fragments, which can be put in order by aligning regions where they overlap.

sequence surprises

During the mid-1990s, one of the companies established by Craig Venter, TIGR, demonstrated the usefulness of genome data by reading the sequences of several bacteria.

Helicobacter pylori is the bacterium that causes stomach ulcers. A stretch of its DNA codes for enzymes that it needs to survive in the strong acid of the stomach; its DNA also codes for other enzymes that help it stick to the stomach lining. Its sequence is being used by physicians and drug companies to develop vaccines and new drugs against ulcers and other diseases caused by *H. pylori*.

Another bacterium, *Deinococcus radiodurans*, is like something from science fiction. It is almost completely resistant to damage by radiation, and also survives starvation, drying out, and just about anything else thrown at it. Its chromosomes are broken into a couple of hundred pieces, but within 12 hours it has stitched them all together, repaired the damage, and is once again replicating. Its DNA sequence is telling scientists about DNA-repair processes, and with a suitably modified genome *D. radiodurans* could be put to work cleaning up radioactive waste.

The NIH turned down TIGR's application for funding in 1995, saying that what he proposed was impossible. At that point the sequence of *H. influenzae* was 90 per cent complete. TIGR published it a couple of months later, the first complete sequence of any free-living organism.

Venter then produced a series of increasingly interesting genomes sequences (see panel, above). At the same time, other sequences were being finished by various groups around the world. In 1995, the heads of the laboratories involved in the Human Genome Project considered the idea that they should publish an unfinished "draft" of the human sequence by 2000, before finishing it properly over the subsequent few years. The US agencies, now headed by Francis Collins, who had replaced James Watson, did not want to be rushed. Collins asked the sequencing centres each to start work on a large chunk of DNA to demonstrate the big improvements in speed and cost needed before they attacked the sequence directly. Venter then forced Collins's hand with another iconoclastic announcement.

the shotgun starts the race

In May 1998, Venter revealed that he had teamed up with the Perkin Elmer Corporation (who made some of the first PCR and sequencing machines) to create a new company – called Celera Genomics – that would sequence the human genome on its own, in three years and for just $300 million (a thousand times cheaper than the forecast for the public project). He also said that he would release data every three months (the publicly funded effort makes daily releases); Celera would make its money by analysing the data and selling the analysis to subscribers. Perkin Elmer's part in this was to provide Venter with 300 new sequencing machines, which were capable of reading more DNA more rapidly and more accurately than anything else available.

The public project responded to the challenge from Venter by announcing a new timetable that would see the genome finished in 2003 – two years ahead of schedule, but two years behind Venter. In October 1999, Celera announced that it had read the first billion bases of the human genome. The Human Genome Project countered that Celera had not released the data for other scientists to evaluate, and in November 1999 went one further by holding a "birthday party" to celebrate its own one billion bases. January 2000 saw Venter claim he had sequenced 90 per cent of the human genome, while

gathering speed
This graph shows the number of bases of human DNA sequenced by the publicly funded Human Genome Project. Improved techniques and faster machines made it possible to read the sequence increasingly quickly.

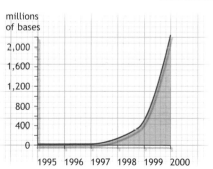

millions of bases

2,000
1,600
1,200
800
400
0

1995 1996 1997 1998 1999 2000

photo finish
The race to read the human genome can be considered a tie, since the public and private sequencing organizations agreed to announce they had finished on the same day. The two draft sequences were published several months later.

PRIVATE

PUBLIC

in March the public project passed the two-billion-base milestone. In the end, the two efforts came in neck and neck, with their joint announcement at the White House in Washington on 26 June 2000. They each had what they called a "working draft" of the human genome, the public consortium being surprised that the Celera version, with more bases, was of similar quality to its own. The two teams appeared to have buried the hatchet, and even talked about a simultaneous publication in one journal. But though the joint announcement was all smiles, there was still a great deal of feuding between the two camps. The Human Genome Project insisted that Celera would not have been able put its many millions of shotgunned fragments in the right order as quickly if it had not had their maps to act as scaffolding. Celera denied that, and argued out that its version was better annotated – meaning that it could give fuller information about the significance of each part of the sequence. But the big fight was over access to the data.

private rights and public goods

One of the best aspects of scientific research is that in theory it is totally transparent. When scientists publish a paper, they are following an unwritten code that insists that they provide all the information needed for another scientist to repeat their work and all the results that another scientist might need to check their conclusions. Access to the outcome of pure research is never restricted. Research laboratories, universities, and commercial concerns might go on to develop and patent applications derived from that information, but the original information is always freely available to other academics. In molecular biology, this has meant that scientists have generally shared the probes, DNA libraries, and sequence data they have accumulated.

Venter bucked that trend as soon as he developed ESTs, which are copies of short pieces of genuine genes, rather than non-coding regions. When he first started working with ESTs, Venter's laboratory was patenting hundreds of sequences a week. Although nobody could tell, from the EST sequence, what the gene that produced it might be for, it seemed a good idea to patent as many as possible. That way, when one proved to be essential in, say, a new drug against obesity, the original patent might prove to be lucrative. (In fact the validity of these patents has not yet been tested; it seems certain that they will be challenged.) Most scientists responded with outrage, and eventually laboratories stopped patenting ESTs. But the episode gave notice of Venter's approach to sequence information.

TIGR is a not-for-profit corporation, but Celera and Human Genome Sciences have shareholders and investors who expect them to make a profit. Sequences are those companies' stock-in-trade. If they give that information away, what do they have left? Venter stated at the outset that he would give everyone "free access" to Celera's

high stakes
Reading a genome requires expensive computers and sequencing machines. This laboratory, at the Wellcome Trust Sanger Institute in England, is funded by a charity, but private companies need to recover their investment by finding ways to market their sequences.

sequence data. For months, scientists and journal editors argued about what exactly "free access" implied. In the end, unable to agree, the two groups published their sequences in the same week but in separate journals.

The publicly funded sequence is available for free, without restriction. Celera's data are available on the company's website, where any researcher can read the sequence. But visitors to the site have to register and agree to specific conditions. Nonprofit workers can search the database and download up to one megabase of sequence a week if they agree not to commercialize or distribute the data. People who want more than this amount have to submit a signed letter from their institution agreeing to Celera's terms. Industry scientists can use the data for free to check the Celera sequence, but if they want to put it to commercial use, they must negotiate an agreement with Celera. Updates to Celera's published sequence also require a fee.

accessing the sequence

The draft sequence of the human genome is stored electronically in various databases, some of which can be accessed free of charge using the internet. As well as the sequence itself, the databases hold several other categories of useful information, some of which can be seen on this display taken from a database called Ensembl.

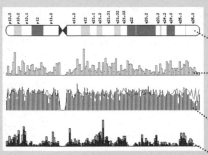

physical map of numbered chromosome regions

numbers of single-nucleotide polymorphisms (single-base differences between sequences)

index of repetition in sequence

numbers of known genes

human genome database

the human genome

As published, the two versions of the human genome sequence represent snapshots of the data as they were in around October 2000. The public sequence is 2.7 billion base pairs long. Chromosomes 21 and 22 (the smallest, apart from the Y chromosome) are finished. Celera's sequence contains 2.65 billion base pairs of connected DNA plus smaller fragments that take the total to 2.9 billion base pairs. None of it is finished to the same standard as the public chromosome 21. Both sequences have about 100,000 gaps. Those gaps include, most notably, the centromeres, which have so far defied all attempts to sequence them completely because the cloning stages essential to the sequencing process have proved to be difficult, if not impossible for these.

Which sequence, if either, is "better"? Almost everyone who knows enough about genomes and sequencing to be able to compare the two sequences is associated with one group or the other. Nevertheless *Nature* and *Science*, the journals that published the public and private sequences respectively, found researchers willing to make the comparison. Both journals concluded that there really was very little difference. By and large, the two sequences match one another and match one of the few maps of genetic markers that was not used to build the sequences.

new information

The biggest surprise was just how few genes there are. Researchers had estimated, based on the number of human proteins they knew about, that there might be

recap

The DNA molecule is packed into structures called **chromosomes**. A typical human cell contains 22 pairs of ordinary chromosomes, plus the X and Y chromosomes (females have two X chromosomes, males have one X and one Y). During the early stages of cell division, each chromosome is duplicated. The two copies join at a region called the centromere, which controls how the chromosomes pair up and divide.

100,000 or so genes. In fact, the number is more like 32,000. That is less than double the number of genes in the nematode worm. But the figure of 32,000 is based on predicted genes; knowing the sequence, and the genetic code that translates bases into amino acids, computer programs are used to figure out which bits of the sequence represent genes. If the computer software currently misses some genes, the final figure, while it will not be lower, could be a lot higher.

However, the lower-than-expected estimate of the number of genes is forcing scientists to look at the way that proteins are made up of different regions, called domains. It no longer seems that one gene necessarily codes for one protein. A single gene can code for several proteins if the protein domains are assembled differently, and one conclusion from the sequence is that on average each human gene can spell out three different proteins. In worms and flies, each gene codes for fewer proteins. It is not quite clear how this happens, but it seems to have something to do with the way that messenger RNAs are assembled when DNA is copied to make proteins. A gene consists of stretches that code for proteins, called exons, interrupted by non-coding stretches, called introns. The way that exons and introns are spliced together to make an mRNA could hold the key to the capacity of a gene to code for more than one protein.

The sequences also reveal that the chromosomes differ in the number of genes they carry. Chromosome 19 has 1,400 in total, over 23 per million bases. By contrast chromosome 13 is

nematode genome

The worm Caenorhabditis elegans *was the first multi-cellular organism to have its genome sequenced. The body of an adult worm consists of exactly 959 cells. The precise location of each of these cells is encoded in the animal's genome, which consists of about 18,000 genes.*

far lighter on genes, with just five per million bases. The sequence also reveals variation in the density of single-nucleotide polymorphisms. Also known as SNPs, these are places where individual people differ by a single letter of code. In some regions, the density is higher than expected, and in others lower. Nobody knows why. Recombination (the swapping of DNA between the members of a chromosome pair during meiosis) is infrequent on chromosome 13, which may or may not have something to do with its low density of genes. But chromosome 12 (in women) and 16 (in men) are recombination hot-spots.

ff The HGP and Cel draft genome assemblies are similar in size, contain comparable numbers of unique sequences . . . and exhibit similar statistics. ??

George Church, geneticist, 2001

The junk DNA is turning up some surprises too. Celera scientists estimate that 40–48 per cent of the sequence consists of repeat sequences, one of the kinds of junk in which a pattern of bases is repeated over and over again. Nearly 10 per cent of the Celera sequence consists of a single kind of repeat, a sequence called Alu, which can consist of up to 280 bases. The Human Genome Project sequence reveals that Alu is clustered in areas rich in genes – perhaps it does have some important part to play, yet to be discovered.

One of the great mysteries, uncovered by the public sequencing effort, is that the human genome shares more than 200 genes with bacteria – but not with nematode worms, fruit flies, or yeast. Some researchers think that an ancient ancestor of vertebrates borrowed some bacteria and their genes in the same way that bacteria are believed to copy antibiotic-resistance genes from other bacteria. But it could equally well have been bacteria that copied vertebrate genes. At the moment, we have no way of knowing the direction of the transfer.

insights from all over

At the same time as sequences were being produced as part of the Human Genome Project, scientists were also learning, and continue to learn, more about the genomes of other organisms. Every sequence that is published reinforces two clear messages. Every genome carries its own surprises. And every genome is related to every other genome. The surprises provide molecular biologists with insights into every aspect of how life works. The relationships, though, provide them with the tools to go fishing for more surprises.

Many, if not most, of the genes found in a particular organism have their counterparts in other organisms. For genes involved in the fundamental processes of life, such as the pathways that make energy available to the cells, the genes are all but identical in every creature studied. Creatures that have been evolving separately for hundreds of millions of years, for example the fruit fly and the human, share genes that are recognisably the same. Indeed, of 1,278 different families of related proteins, only 94 are unique to vertebrates (the group of animals with backbones that includes humans). To put it another way, more than 90 per cent of protein families had evolved before the first vertebrates, more than 500 million years ago.

adult fruit fly

fly chromosome

gene

spinal chord

mouse embryo

mouse chromosome

homoetic genes
The development of some animals' body parts is controlled by homoetic genes. These genes and the corresponding body parts are shown here in color; other genes are shown in grey. The homoetic genes and the parts of the body they control are arranged in the same order. In fruit flies, for example, the genes that control development of the head are followed by genes for the thorax and then for the abdomen. A similar pattern can be seen in other animals, such as the mouse, suggesting that the homoetic genes have a common origin.

There are genes called homoetic genes that control the body pattern of the fruit fly, which specify, for example, that an appendage will be an antenna if it is on the head but a leg if it is on one of the fly's body segments.

leg produced in place of antenna

the wrong instructions
When a homoetic gene is mutated, the results can be bizarre. For example, if the gene controlling the development of a fruit fly's antennae is mutated, the fly may have legs where its antennae should be.

Every one of these genes contains an identical 180-letter sequence called the homeobox, which is odd. If it is the same in every gene, it cannot be saying anything about whether to grow an antenna or a leg. As the British author Matt Ridley has noted, "all electrical appliances have plugs, but you cannot tell a toaster from a lamp by looking at the plug". In fact the analogy is acute; the homeobox seems to code for a part of the protein that allows the protein to plug into the DNA somewhere else. The protein latches onto a target sequence of the DNA and stops it being copied. Proteins like this switch the genes on or off.

Acting on a hunch, other scientists used the fruit fly's homeobox sequence to look for similar sequences in the genome of a frog. Somewhat to their surprise, the homoetic genes turned up in the frog. And the mouse. And the human. At some very basic levels, all creatures are indeed alike.

Practically every scientific paper on genes uses these similarities to track down connections. For example, the disease called narcolepsy afflicts about one in 2,000 adults. Sufferers experience various problems, including a tendency to fall asleep during the daytime. In humans, the disease is not inherited in any simple fashion, but in dogs there is a single recessive gene that gives rise to narcolepsy. Scientists at Stanford University in the US found the gene in dogs; it coded for a receptor,

dog tired
Narcolepsy affects various animals. Scientists have identified a recessive gene that causes the condition in dogs.

a kind of aerial that sticks out of a cell and allows it to receive signals from other cells, though they did not know what the signalling molecule was. Other scientists in Texas discovered that the absence of a particular signalling molecule in mice made the mice narcoleptic. With the receptor sequence from dogs, and the signal from mice, researchers went looking in the human genome, and found that people who suffer from narcolepsy have the right receptor gene, but for some reason, still unknown, do not make the signal. This kind of cross-species study, and the information that it reveals, is enabling pharmaceutical companies to come up with better drugs to treat conditions that until now have been very difficult to deal with.

the rise of bioinformatics

Although work on finishing the sequence continues, a substantial effort is also being applied to making sense of the data. And just as sequencing was utterly dependent on technical advances – the development of automated sequencing machines that could do in a matter of hours what used to take a person years – new advances require computers and a new breed of unsung heroes, called bioinformaticians.

A piece of sequence, on its own, tells you almost nothing. Does it code for a gene? Is it related to another sequence? If it does code, what kind of protein might it make? Has someone else already discovered what that kind of protein does? These are easy questions to ask, but there is now so much data that they are almost impossible to answer without powerful computers running sophisticated software. Designing this software and putting it to use is the work of bioinformaticians.

Whenever a group of researchers in the public project finishes a piece of sequence, they deposit the details in

the finishing touches
The human genome sequence still contains gaps and errors. Completing the Human Genome Project will be a long and difficult task because the parts of the genome that remain unread are among the most difficult of all to sequence.

one of a few databases. Results are regularly shared between the databases, so each of them has all the available sequence information. And each database provides software tools to allow researchers to interrogate it, each tool specialized to answer different questions. This is where the cutting edge is now: making it easier for researchers to come to grips with the information and detect patterns in the data. The 3 billion bases of the human genome are just part of the total volume of data. There are the genomes of all the other organisms that have been completely sequenced, as well as bits and pieces from the genomes of hundreds of other plants, animals, and microorganisms that have been of particular interest to particular researchers. Each bit of information added makes all the rest more valuable, and the software tools that enable scientists to mine meaning from the endless strings of As, Ts, Cs, and Gs are the most valuable of all.

❝ Each new round of press conferences announcing that the human genome has been sequenced saps the morale of those who must come to work each day actually to do what they read in the newspapers has already been done. ❞

Maynard V. Olson, geneticist, 2001

what now?

The business of "finishing" the human genome will continue, but only in the public project. This involves such tasks as filling in the gaps and repeating parts of the sequencing process to eliminate errors. It will, however, take enormous dedication to carry on. After all, the glamorous work – finding out what all those genes do and how to make use of them – will be elsewhere. Some of the sequence will never be finished. Regions such as the centromeres may defy all attempts to read them in detail. But even the first draft enables us to glimpse the enormous potential available in the information.

pitfalls and promises

The human genome sequence is morally neutral, ethically unimpeachable. That is one sense in which it is not like a book. Books can be immoral and unethical, because words carry meaning, which can influence how people think and behave. The sequence carries no meaning; it is how people use it that matters. It could bring massive benefits, massive problems, or both. Most of the things that worry ordinary people about the human genome – cloning, designer babies, genetic discrimination – do not need the sequence, and were happening before anyway. They provoked questions before the sequence was published, which the sequence has not answered. But the very fact of having a draft sequence, and the world-wide attention it has attracted, has propelled those questions back into the limelight. It is no longer wise to ignore them and hope they will go away.

" We are creating a world in which it will be imperative for each individual person to have sufficient scientific literacy to understand the new riches of knowledge, so that we can apply them wisely. "

David Baltimore, biologist, 2001

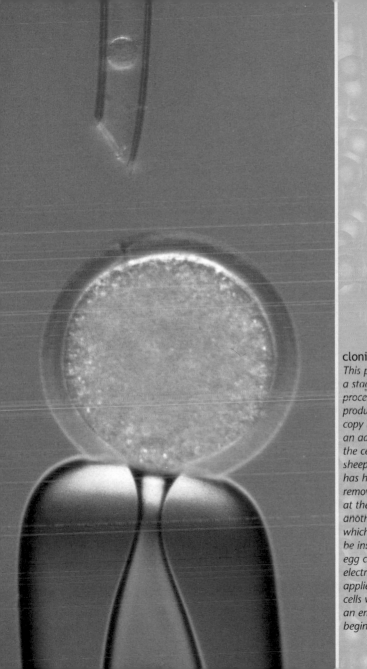

cloning
This picture shows a stage in the process used to produce an exact copy (or clone) of an adult animal. At the center is a sheep egg cell that has had its nucleus removed. The tube at the top holds another sheep cell, which is about to be inserted into the egg cell. When an electric charge is applied, the two cells will fuse and an embryo will begin to develop.

carbon-copy clones

Dolly is probably the most famous sheep in the world, because she is a clone. When people talk of a clone they can mean many things, but essentially a clone is an exact copy. A cloned piece of DNA is a copy of another piece of DNA. A named rosebush or potato is also a clone (though we never seem to call them that). In 1997, scientists working at the Roslin Institute, in Scotland, announced that they had created Dolly using a technique known as somatic cell nuclear transfer (see panel, below). This is what people usually mean when they talk about making a

animal cloning

The technique used to clone Dolly required two donors. One donor (Dolly's "mother") provided a cell nucleus, complete with its DNA. Another donated an egg cell in which the nucleus was destroyed. Dolly's nucleus was then inserted into the donor egg cell, which was implanted in the womb of another ewe. This ewe gave birth to Dolly. In most cells, only the genes needed by that cell are "switched on." But by starving the cell from which the nucleus came of nutrients, its development was halted when all its genes were still active.

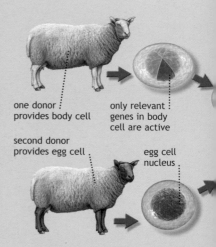

one donor provides body cell

only relevant genes in body cell are active

second donor provides egg cell

egg cell nucleus

clone. The other way of making this kind of clone is known as blastomere separation. At a very early stage in the development of an animal, when it is still a ball of 8 or 16 cells, it is possible to separate each of those cells and have them all develop into genetically identical individuals. Indeed, that is exactly how identical twins come about in nature. Blastomere separation goes on all the time, especially in animal breeding. It no longer excites much interest or debate.

The important thing about Dolly is not that she is a clone; it is that she was derived from an adult cell. A single fertilized egg cell has all the genes needed to build all the different kinds of cells a complete body contains, but as the cells multiply they become increasingly differentiated. An embryo can make any kind of cell. But when a differentiated skin cell divides, for example, it can only produce more skin cells. It still has all the genes, but

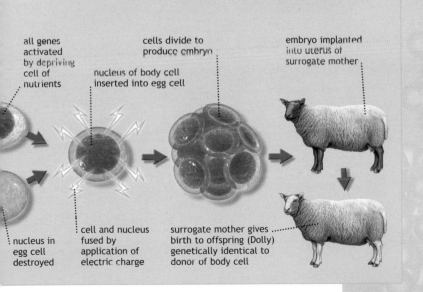

all genes activated by depriving cell of nutrients

cells divide to produce embryo

nucleus of body cell inserted into egg cell

embryo implanted into uterus of surrogate mother

nucleus in egg cell destroyed

cell and nucleus fused by application of electric charge

surrogate mother gives birth to offspring (Dolly) genetically identical to donor of body cell

all except those needed to be skin are "switched off." Dolly proves that it is possible to take a cell from an adult animal and switch all the genes back on.

human clones

Why clone humans? Cloning could allow a totally infertile couple to replicate one or both parents. Grieving parents might resurrect a dying child. Homosexual couples could create offspring to nurture. A sports fan could raise his hero again. You can think of other possibilities. Undoubtedly some would give great satisfaction, but they have to be set against the possible drawbacks. To create Dolly, the Roslin scientists transferred 277 nuclei, which gave them 29 normal-looking embryos, which resulted in one live birth. Other animal cloners have boosted the success rate, but the ratio of births to implants is still only about one in 20. Cloned animals, furthermore, seem to have all sorts of problems that may not be apparent at birth. Cloned mice become very obese, and cloned calves have difficulty breathing and malfunctioning livers. Dolly may have been luckier than we know.

Some of these problems may be ironed out in time, but is it right to even attempt to clone humans before they are? Even when animals are being routinely and safely cloned (and it is hard to think of good reasons why they need be) will it be right to make a human being with a specific genetic makeup? There are other ways of dealing with childlessness and bereavement that, far from assailing human dignity, enhance it.

> **" What will happen when the adolescent clone of Mommy becomes the spitting image of the woman with whom Daddy once fell in love? In case of divorce, will Mommy still love the clone of Daddy, even though she can no longer stand the sight of Daddy himself? "**
>
> Leon R. Kass, bioethicist, 2001

copying people
A fear raised by new cloning technology is that it could be used to make multiple copies of people. With present technology, making each clone would require a surrogate mother. And as the clones grew, they would become different, each shaped by its unique environment.

treating disease

One of the things that most excites people about genetic information is its medical potential. We all want to live long, healthy lives, and understanding how genetic information contributes to disease may help fulfill that desire.

At the moment, a lot of hope is pinned on stem cells. During development, as cells become increasingly dedicated to specific functions, they reach a stage called the stem cell. These cells are capable of becoming several different kinds of more specialized cell, but not all. The embryo contains powerful stem cells, capable of becoming many different kinds of cell, and these could be useful in treating some diseases.

Parkinson's disease, for example, is caused by a lack of certain cells in the brain, while diabetes is due to a lack of specialized pancreatic cells. It might be possible to cure these kinds of diseases by transplanting the appropriate kind of stem cell. At the moment, there are two basic approaches to stem cell research. Some groups are trying to work out whether it is possible to reset the switches on a more developed cell to turn it back into a stem cell. Others are working with stem cells harvested from embryos.

The ethical dilemma seems to revolve around the source of stem cells. Some people argue that all research on stem cells is unacceptable because they can only be obtained from embryos, which, they claim, are human beings. But fertility clinics routinely make more embryos than they put back into women seeking in-vitro fertilization. If these extra embryos can provide stem cells that will help people with

stem cells
These stem cells are from human bone marrow. They have the potential to become red blood cells or one of several kinds of white blood cell. Blood cells degrade relatively quickly and are replaced almost continuously.

diseases, surely that is more useful than destroying them? And if they are neither destroyed nor made useful, what is the point of storing them forever? Limiting research to stem cells that already exist seems to be fudging the decision.

screening for survival

Neither cloning nor stem-cell research require the full genetic sequence. Nor does genetic screening, though the sequence can make it more useful and effective. Genetic screening takes many forms, but essentially each of them reveals whether a person carries a particular piece of genetic code. The problem is what to do with that information.

Some diseases are caused by mutation of a single gene and are inherited very simply. If that is the case, it can be helpful, especially for couples planning to have children, to know whether they have the gene. Cystic fibrosis, for example, and sickle-cell anemia, are harmful only when a person has two copies of the mutation. If a couple each has one copy of a mutation – which neither of them would know without screening – each of their children has a one-in-four chance of having the full-blown disease. Often the first intimation a couple has that they are both carriers is the birth of an affected child. In that case, screening the embryo in a subsequent pregnancy can give them the information they need to make a decision that suits them.

Some groups go further. Ashkenazi Jews tend to carry a mutation that causes a debilitating and fatal disease called Tay–Sachs disease. An organization in the United States arranges for Ashkenazi schoolchildren to have genetic tests for

recap

A genetic **mutation** is a change to DNA caused by a copying error or environmental factors. Not all mutations are harmful, and many have no effect, but they can sometimes cause disease.

sickle cells
Sickle-cell anemia is caused by a mutation that is relatively easy to detect by screening. The mutation leads to production of an abnormal form of the protein hemoglobin, which causes some red blood cells to be distorted from their normal rounded shape into a distinctive sickle shape.

screening for sex

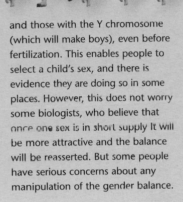

Throughout history, some parents have wanted to choose the gender of their children. In mammals, gender is controlled by a gene on the Y chromosome called the SRY gene. Having the SRY gene makes an individual male; not having it makes them female. The discovery of this gene, along with techniques for DNA amplification, makes it possible to determine the eventual sex of a fertilized embryo once it has divided a few times. Later-stage fetuses can be sexed in the same way. It is also possible to sort sperm into those with the X chromosome (which will make girls) and those with the Y chromosome (which will make boys), even before fertilization. This enables people to select a child's sex, and there is evidence they are doing so in some places. However, this does not worry some biologists, who believe that once one sex is in short supply it will be more attractive and the balance will be reasserted. But some people have serious concerns about any manipulation of the gender balance.

Tay–Sachs disease and cystic fibrosis. When a couple are thinking of getting married (often at the suggestion of matchmakers), they phone the organization, which matches the anonymous numbers each was given when tested against the results. If both are carriers, the organization advises against the marriage. As a result, these genetic diseases have almost vanished from the Ashkenazi population in the US. Couples who do not enter arranged marriages in this way can be screened, but if they learn that they are carriers they will face some difficult decisions: they might choose to remain childless; if not, they could conceive and either run a one-in-four risk that the child will have the disease or screen the embryo and selectively abort the pregnancy.

Unfortunately, diseases that are simply dealt with are in the minority. Most raise complicated questions. For many, although screening offers a diagnosis, no treatment is available. Huntington's chorea, which afflicts people in middle age (after they may have had children themselves), is a disease with no cure. There is a genetic test available, the results of which are essentially either a death sentence or a free pardon. Some people do avail themselves of the test, which has to be handled with extraordinary sensitivity and insight, and not all sufferers experience the depression or suicide that had been predicted.

Most genetic tests are much less clear cut. One kind of breast cancer is linked to two different genes. But not all women with the mutation develop the disease; there are obviously environmental factors that we know little about. Nevertheless, knowing that she has a genetic predisposition, a woman could make choices that might make the disease less likely to develop. Preventative removal of the breasts is not a choice on offer, although some women do have it. But detection of another gene, linked with cancer of the thyroid, may be followed by removal of the thyroid, whose hormones are easily replaced with pills.

Screening can also be used to help a sick person directly. A child may have a serious disease that can be cured by a transplant, but no donor is available. The child's parents may want another child anyway. If they are lucky, that child may have the right genes to help its sibling. But why not make certain? Why not fertilize several eggs outside the body, test to see which have the correct genetic constitution, and then implant only those in the mother? It has already happened, more than once.

genetic apartheid

The big danger at present is that our ability to screen for genes is running far ahead of our ability to do anything about what we discover, and that has been accentuated by

Huntington's chorea
Huntington's chorea, the disease that killed the American folk singer Woody Guthrie (background, right), is associated with a genetic mutation. Screening can be used to identify the mutation, and can also give a very precise indication of the age at which symptoms will develop. However, at present there is no cure for the disease.

protective diseases

Some recessive mutations are very common among particular groups of people. Having two mutated genes causes the disease, which is usually fatal, but the mutation survives in carriers because it protects against some other disease. Sickle-cell anemia, common in black people from Africa, defends carriers from malaria. Cystic fibrosis carriers are protected from typhoid. And Tay–Sachs disease seems to keep carriers safe from tuberculosis. Where those diseases are common, being a carrier can be a good thing, despite the sacrifice of losing one child in four. But when there are other ways to deal with the disease, the mutation is no longer necessary. Natural selection will eventually get rid of it.

the decoding of the genome. In the course of sequencing the genome, thousands of markers were located, many of them single nucleotide polymorphisms, which are particularly easy to find. This makes searching for even the most tenuous of links relatively simple. But in the absence of any therapy – or even good advice – what good is it to know that you have a better than average chance of developing, say, bowel cancer?

One answer is that it can make use of medical resources more efficient. Many cancers are most effectively dealt with if caught early. Someone with a family history of bowel cancer used to have no option but to submit to frequent examinations. Knowing that you do not carry the gene enables you to make a rational decision to be examined less often. Of course, you may still develop cancer of the bowel or some other organ; mutations, possibly the result of cosmic rays or environmental chemicals, can still cause cancer even if you have not inherited a genetic predisposition.

huntington's?

parkinson's?

cancer?

ignorance is bliss?

In some cases, genetic screening can provide people with information about their predisposition to disease that is of real use. However, in other situations, particularly where treatment options are limited, screening may present people with difficult decisions.

But what if you are screened, and as a result an insurance company raises your premium or, worse, refuses to cover you at all? What if a prospective employer refuses you a job? This has caused concern, and legislative attempts have been made to prevent companies both using genetic information and passing that information on to others.

> **" Like racism, sexism, and other forms of prejudice, genetic discrimination devalues diversity, squanders potential, and ignores achievement. "**
>
> US Department of Health and Human Services, 1997

For employment, it is hard to justify genetic discrimination, an issue addressed by courts in the US when companies refused to hire people with one copy of the sickle-cell gene. But in the case of insurance, the picture is cloudier.

Insurers have always taken an avid interest in anything that affects your chances of survival. That is why they offer lower premiums to people who don't smoke, for example. Application forms commonly ask about any ailments your parents may have suffered from; if both your parents died young of heart disease, you are likely to, too, and you may be asked to pay a higher life-insurance premium. When applied to life insurance, it is hard to know how genetic screening – which is more accurate than life history, and could just conceivably result in lower premiums for some people – could be a bad thing.

all in the genes?
Humans are much more than the product of their genes, but following the publication of the draft sequence of the human genome, we now know more than we did before about how alike we are to each other and about what makes us different from other species.

where to now ?

The use of sequence information to treat diseases is advancing rapidly. Drug companies scan the sequence for particular messages that they might mimic. If a signal is missing, the mimic can supply it; if there is too much, then the mimic might switch off the gene or block the receptor. The sequence also helps diagnosis. It is possible to put microspots of thousands of pieces of DNA on a glass slide. If that is washed with a preparation of the DNA of a patient's cells, active genes will stick to the microspots, giving a kind of fingerprint of the genes in that kind of cell. This has enabled doctors to separate, for example, one kind of leukaemia into two classes, only one of which responds well to treatment. With the new knowledge provided by the sequence, it should be possible to devise new treatments to help those who previously had no hope.

It is impossible in a book of this kind to cover all the possible uses to which genetic information might be put. Medical reasons, or "freedom" will always be cited to justify certain kinds of practice, some of which are perfectly reasonable while others make us feel uneasy. There can be no hard and fast rules about what to permit and what to forbid. But it is as well to remember that some of the medical suggestions are effectively shortcuts to solutions for problems that might be solved in other ways. And there are already plenty of freedoms that society chooses to curtail. There are fine distinctions to be made; an educated public and well-informed politicians are the best guarantee that they will not be made foolishly.

glossary

allele
One of two or more forms of a gene. In organisms that reproduce sexually, an individual inherits one allele from each parent. If these two alleles are different, the one that is expressed in the offspring is referred to as the dominant allele; the other is termed the recessive allele.

amino acid
A nitrogen-containing compound. Amino acids are the basic building blocks of proteins.

amplification
The multiplication of a particular stretch of DNA, either chemically (*see* **polymerase chain reaction**) or biologically (*see* **cloning vector**).

base
A basic component of nucleic acids. There are two kinds: pyrimidines and purines. DNA contains the pyrimidines thymine and cytosine, and the purines adenine and guanine. In a DNA helix, adenine pairs with thymine, and guanine pairs with cytosine. The same bases occur in RNA, except that thymine is replaced by uracil.

bioinformatics
The use of computer hardware and software to store and interpret data derived from sequencing research.

blastomere separation
The separation of the cells of a fertilized egg early in the development of an embryo. Each cell may develop into a whole organism. All the organisms from a single blastomere will be genetically identical clones.

centromere
The part of a chromosome where the two members of a chromosome pair come together during cell division.

chromosome
A mixed package of DNA and protein. Chromosomes (with the exception of the sex chromsomes) occur in pairs, with one in each pair coming from each parent. The sex chromosomes are known as X and Y: female mammals have two X chromosomes; males have one X and one Y.

clone
An identical copy. Cloned DNA has been inserted into a bacterium or virus, which makes several copies of the DNA. Cloned organisms contain identical DNA. Clones may be natural (such as identical twins) or artificial.

codon
Also known as a triplet, a three-letter "word" of the genetic sequence, where each letter is one of the four bases. Each codon either represents one of 20 amino acids found in proteins or an instruction to start or stop assembling a protein.

complementary DNA (cDNA)
DNA that has been copied from a piece of mRNA.

complement
A single strand of DNA that corresponds uniquely to another single strand according to the rules of base pairing.

contig
A long stretch of DNA sequence, constructed by assembling the sequence read from smaller pieces of DNA.

crossing over
see **recombination**

DNA
Deoxyribonucleic acid. DNA is a nucleic acid in which the bases occur in pairs. The sequence of base-pairs encodes the information needed to manufacture proteins and, in turn, to produce a complete organism. For this reason, DNA is regarded as the essential molecule of heredity.

dominant allele
see **allele**

expressed sequence tag
cDNA made from mRNA, and therefore most likely from a gene rather than a noncoding region.

gel electrophoresis
A technique for separating a mixture of molecules, such as pieces of DNA or proteins, by putting them in an electrically charged gel.

gene
The unit of inheritance. There are several ways of defining a gene more precisely. In classical genetics, a gene is a hereditary factor that controls a characteristic of an organism. In molecular terms, a gene is a sequence of DNA that contains the code to make a protein.

genetic map
see **linkage map**

genome
The entire genetic message of an organism.

homoetic gene
A gene controlling the development of a large unit of the body, such as the legs or wings.

linkage
The relationship between two genes, which are inherited together more often than expected.

linkage map
A map of the position of genes relative to one another, drawn up by measuring how often different genes are separated from one another during recombination.

junk DNA
A presumptuous label for a stretch of DNA whose function scientists do not fully understand. Junk DNA includes repetitive sequences and sequences called introns, which divide two meaningful sections, or exons, of a gene.

marker
Any portion of DNA that can be followed from one generation to the next. A marker may be a whole gene that has a visible effect, or a single letter of the code that differs between individuals.

meiosis
A process of cell division in which the amount of genetic material in the daughter cells is half that contained in the parent cells. As a result of recombination during meiosis, each daughter cell has a unique set of genes. Meiosis produces the egg and sperm cells that are involved in sexual reproduction.

mitosis
A process of cell division that produces daughter cells having an identical set of genes to the parent cells. Mitosis produces the new cells an organism needs for growth and to replace worn or damaged cells.

mRNA
see **RNA**

mutation
A change to DNA. Most mutations in genes are silent, in that they do not change the protein coded for by the gene.

nucleic acid
A long molecule made up of alternating bases and sugar-phosphate molecules.

nucleus
A discrete body within a complex cell. The nucleus contains the DNA in the form of chromosomes.

physical map
A map of absolute positions of genes, on which distances between genes can be measured in base pairs. A map produced by reading the actual sequence of genes is an example of a physical map.

polymerase chain reaction (PCR)
A fast and accurate chemical method for identifying and amplifying particular stretches of DNA.

polymorphism
Any difference between the genes of two individuals. A single-nucleotide polymorphism (SNP) is a difference in a single base. Not all SNPs produce differences in an organism's offspring. Other polymorphisms produce large, visible effects, such as the difference between brown and blue eyes.

protein

A long molecule assembled from amino acids. Proteins are the workhorses of the cell: they carry messages within and between cells, enable the chemical reactions essential for life, and form many of the structural components of living things.

protein domain

A region of a protein, for example, a receptor for another protein or a type of "anchor" that keeps the protein in a particular place.

recessive allele

see **allele**

recognition site

A stretch of DNA that is recognized by a protein. The protein may be a switch that toggles a gene on and off, or it may break the DNA into fragments.

recombinant clone

DNA (or a whole organism) made by deliberately combining DNA from two different organisms, often from different species.

recombination

The exchange of DNA between the two members of a chromosome pair during meiosis. Recombination shuffles the DNA from mother and father to create an entirely new individual.

restriction enzyme

A protein (usually isolated from a bacterium) that cuts DNA at a particular

sequence. Some restriction enzymes, known as rare cutter enzymes, recognize a long sequence, eight bases or more in length. These long sequences occur relatively infrequently, so the enzymes break the DNA into large fragments.

RNA

Ribonucleic acid. RNA differs from DNA in the sugar (ribose instead of deoxyribose) and in one of the bases (uracil instead of thymine). Genes on DNA are transcribed into messenger RNA (mRNA), which leaves the nucleus to guide the manufacture of protein in the cytoplasm. Small RNA molecules (tRNA) transfer amino acids to the growing protein chain. Some viruses store their genetic information as RNA (rather than as DNA).

shotgun sequence

A sequence created by breaking DNA into fragments, reading each fragment, and then using computer software to assemble the fragments into a single sequence.

single-nucleotide polymorphism

see polymorphism

sticky end

A single strand projecting beyond a piece of double-stranded DNA. Sticky ends produced by cutting DNA with the same restriction enzyme can be used to recombine DNA from different sources.

translation

A process in which the genetic code in a DNA molecule is used to manufacture a protein.

tRNA

see **RNA**

restriction-fragment-length polymorphism

A polymorphism detected because a mutation changes the recognition site of a restriction enzyme. The enzyme creates fragments of different lengths in DNA from different individuals.

stem cell

A cell that has not fully differentiated, and which can become any one of several types of cell.

somatic cell nuclear transfer

A cloning technique that involves moving the nucleus from a somatic (body) cell into a fertilized egg. The egg may then develop into an individual that is a genetically identical clone of the donor of the somatic cell nucleus.

variable-number tandem repeat

A polymorphism caused by a "stutter" in DNA, which causes the length of a repeated sequence to differ between individuals.

vector

An organism (usually a bacterium or virus) used to carry and multiply a piece of DNA from another organism.

index

Further reading

Man-Made Life, Jeremy Cherfas, Pantheon Books, 1982

The Sun, the Genome, and the Internet, Freeman J. Dyson, Oxford University Press Inc., 1999

The Eighth Day of Creation, Horace Freeland Judson, Cold Spring Harbor Laboratory Press, 1996

Genome, Matt Ridley, Fourth Estate, 1999

Useful web addresses

The Celera genome sequence
public.celera.com/index.cfm

The Human Genome Project sequence
www.ncbi.nlm.nih.gov/genome/guide/human

The Genome News Network
gnn.tigr.org/main.shtml

MendelWeb
www.nctspace.org/MendelWeb/

Nature
www.nature.com/nsu/

The Wellcome Trust Sanger Institute
www.yourgenome.org

Acknowledgments

Dorling Kindersley would like to thank Don Powell and Hazel Richardson for editorial advice and assistance. The index was prepared by Richard Raper and Simon Field of Indexing Specialists.

Jacket design
Nathalie Godwin

Picture librarian
Richard Dabb

Picture credits
Corbis: Bettman 19cr. Hulton Archive: Frank Driggs Collection 62c. King's College London: 23tr. National Library of Medicine: 7br, 18bc, 34c. Nobel Foundation: 33cra. Oxford Scientific Films: Paul Franklin 9tl. Science Photo Library: 10cl, 13c, 28-29c,r; Addenbrokes Hospital, Dept. of Clinical Cytogenetics, 40ca; A. Barrington Brown 23bc; Biophoto Associates 3l; Dr. Jeremy Burgess 19bc; CNRI 5cl, 15br, 18bc; Ken Edward 1c; Eye of Science 59tr; Klaus Guldbrandsen 35tc; Jackie Lwein, Royal Free Hospital 60bc; Dr Gopal Murti 4-5c; Alfred Pasieka 54-55c,r; Philippe Plailly 40cb; Plailly, Eurelios 57tc; W. A. Ritchie, Roslin Institute, Eurelios 55l. David Scharf 51tr; Sincair Stammers 48bl; Andrew Syred 22tc, 28-29c,r; Jean-Yves Sgro: Institute of Molecular Virology, University of Wisconsin-Madison 33tl-c. Sweden Post Stamps: 22cl. Corbis Stock Market: John Martin, 98 44tl. Stone / Getty Images: Elie Bernager 64-65bc,bl. The Wellcome Institute Library, London: 41tc. The Wellcome Trust, Sanger Institute, Cambridge: 45bl; EMBL, EBI 46bl. The Whitworth Art Gallery, The University Of Manchester: 6bc. Jacket: Science Photo Library: Ken Edward

Every effort has been made to trace the copyright holders.
The publisher apologizes for any unintentional omissions and would be pleased, in such cases, to place an acknowledgment in future editions of this book.

All other images © Dorling Kindersley.
For further information see: **www.dkimages.com**